Frogs and TOADS

Rebecca Woodbury, Ph.D., M.Ed.

I0059385

Gravitas Publications Inc.

Frogs and
TOADS

Illustrations: Janet Moneymaker

Copyright © 2024 by Rebecca Woodbury, Ph.D., M.Ed.

All rights reserved. No part of this publication may be reproduced, stored in a retrieval system, or transmitted, in any form or by any means, electronic, mechanical, photocopying, recording, or otherwise, without prior written permission from the publisher. No part of this book may be reproduced in any manner whatsoever without written permission.

Frogs and Toads
ISBN 978-1-950415-65-6

Published by Gravitas Publications Inc.
Imprint: Real Science-4-Kids
www.gravitaspublications.com
www.realscience4kids.com

RS4K

Photo credits: Cover and Title Page: Cinder 1280, CC BY A 4.0 International; Above, Bnski-CC BY SA 3.0; P. 5. – 1. David Mark from Pixabay; 2. Alexas Fotos from Pixabay; P. 7. André Karwath aka Aka, CC BY SA 2.5; P. 9. – 1. By David Clode on Unsplash; 2. By Brian Gratwicke-CC BY SA 2.0; 3.By Michael Gäbler-CC BY SA 3.0; 4. By Cephas, CC BY SA 3.0; P. 13. – 1. Visviva, Public Domain; 2. Bnski, CC BY SA 3.0; 3. Rushenb, CC BY SA 4.0; 4. JurriaanH, CC BY SA 3.0; P. 17. – 1. Iwona Grabska-CC BY SA 3.0; 2. Viridiflavus-CC BY SA 3.0; P. 21. By Cathy Keifer, AdobeStock

Frogs are animals that
live on land and in water.

Some frogs start out in the water and then live only on land as adults. Other frogs continue to live both on land and in water.

We like the land!

I like both water and land.

Toads start out in water but live only on land as adults.

Pretty!

I want to dye my hair with red spots!

Frogs and toads are called **amphibians.** The word **amphibian** means "two lives." Frogs and toads live at least part of their lives on land and in water.

Frogs and toads belong to a big group of animals that scientists call **Anura**.

Frogs and toads look different from each other.

Frogs have long legs.
Toads have shorter legs.

Frogs have smooth skin.
Toads have bumpy skin.

Frogs and toads come in different sizes and colors.

Toad

Toad

Frog

Frog

Frogs and toads have a **life cycle.**

A **cycle** is a series of steps that repeat.

Frog Life Cycle

Eggs hatch into tadpoles.

Adult frogs lay eggs.

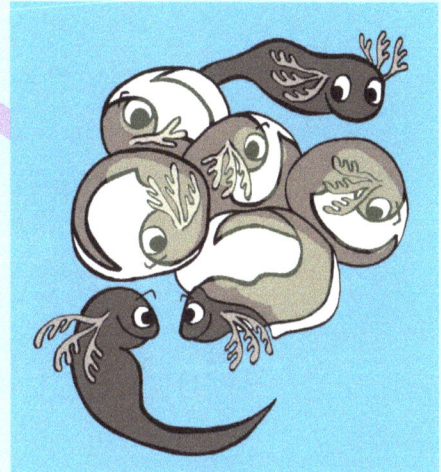

Tadpoles grow into adult frogs.

Frogs begin as eggs in water.
The eggs hatch into **tadpoles**
that live in the water and grow
into adult frogs.

I am an adult.

Are you sure?

Adult frogs lay more eggs and the life cycle repeats.

Do mice lay eggs?

No, silly! Mice have babies!

Frogs and toads are important.
They eat many insects and are
food for other animals.

A frog uses its sticky tongue to catch insects!

How to say science words

amphibian (am-FI-bee-uhn)

Anura (uh-NOO-ruh)

frog (FRAHG)

life cycle (LIYF-SIY-kuhl)

science (SIY-uhns)

scientist (SIY-uhn-tist)

tadpole (TAD-pohl)

toad (TOHD)

www.ingramcontent.com/pod-product-compliance
Lightning Source LLC
Chambersburg PA
CBHW040150200326
41520CB00028B/7551